D1644056

WATER POWER

POLLY GOODMAN

HODDER
Wayland

an imprint of Hodder Children's Books

A L I S
1682246

LOOKING AT ENERGY

Other titles in the series

Fossil Fuels · Geothermal and Bio-energy
Nuclear Power · Solar Power · Wind Power

For more information on this series and other Hodder Wayland titles, go to www.hodderwayland.co.uk

This book is a simplified version of the title 'Water Power' in Hodder Wayland's 'Energy Forever?' series.

Language level consultant: Norah Granger
Editor: Belinda Hollyer   Designer: Jane Hawkins

Text copyright ©2001 Hodder Wayland
Volume copyright ©2001 Hodder Wayland

First published in 2001 by Hodder Wayland
an imprint of Children's books.

This paperback edition published in 2005

All rights reserved. Apart from any use permitted under UK copyright law, this publication may only be reproduced, stored or transmitted, in any form, or by any means with prior permission in writing of the publishers or in the case of reprographic production in accordance with the terms of licences issued by the Copyright Licensing Agency.

British Library Cataloguing in Publication Data
Goodman, Polly
Water power - (Looking at energy)
1.Water power - Juvenile literature
I.Title
333.9'14
ISBN 0 7502 4721 5

Printed in China by WKT

Hodder Children's Books
A division of Hodder Headline Limited
338 Euston Road, London NW1 3BH

ABERDEENSHIRE LIBRARY AND
INFORMATION SERVICE

Goodman, Polly

Water power /
Polly Goodman

CAW   347172

H333.914   J333.
                      914

1682246   EXH

Picture Acknowledgments

Cover: main picture Ecoscene/Mike Maidment, waterwheel The Stockmarket. The Stockmarket: title page, 9, 20-21 (Ballantyne), 25 (Rose). US Department of Energy: pages 4-5. Olë Steen Hansen, Denmark: page 5. Lionheart Books: pages 6-7. Eye Ubiquitous: pages 7 (NASA), title page, 16-17 (Brian Pickering), 41 (Steve Lindridge), 43 (Davy Bold). Ecoscene: pages 8-9 (Andrew Brown), 10 (Chris Knapton), 12-13 (Erik Shaffer), 13 (Andrew Brown), 14-15 (Richard Glover), 26 (Joan Creed), 32 (Rob Nichol), 36 (Nick Hawkes), 40 (M.Jones). Mary Evans Picture Library: pages 17, 19 (both), 20. AEA Technology: pages 22, 30, 30-31, 34, 38-39. Samfoto: pages 28-29 (Jon Arne Sieter), 29 (Morten Loberg). Kvaernar Brug, Norway: page 32 (both). Bruce Coleman Ltd. (Goldon Langsbury) page 35. Frank Lane Picture Agency/ McCutcheon, Alaska, USA: pages 42-43.

CONTENTS

WHAT IS WATER POWER?

Water power is a source of energy. It can be used to run machines and make electricity. Water power is a clean source of energy that can be used again and again.

A dam on the Columbia river, in the USA, used to produce electricity. ▼

Water power has been used for thousands of years. In the nineteenth century, steam power took over from water power. Then in the twentieth century, fossil fuels took over as the main source of energy.

Now water power is being used again. Fossil fuels are running out, and in any case they pollute the environment with harmful waste. Scientists and engineers are looking for ways to use water power instead of fossil fuels.

▲ A old type of water wheel in Denmark. It is used to grind grain to make flour.

FACTFILE

About 6 per cent of energy used in the world is electricity made by water power. Almost all of it is made by hydroelectric power stations.

WATER EVERYWHERE

▲ Niagara Falls on the border between the USA and Canada. The Falls are used to make electricity – enough for a large city.

The water cycle

Water covers three-quarters of the earth's surface. It constantly moves between the oceans, rivers and clouds in a natural cycle, called the water cycle.

The sun heats the water in oceans and rivers. The water evaporates and rises up into the atmosphere. When winds carry the wet air to cooler areas, it condenses and forms tiny water droplets. The water droplets form clouds.

Then the water in clouds falls as rain. The water flows down to the rivers, lakes and oceans, and the cycle begins again.

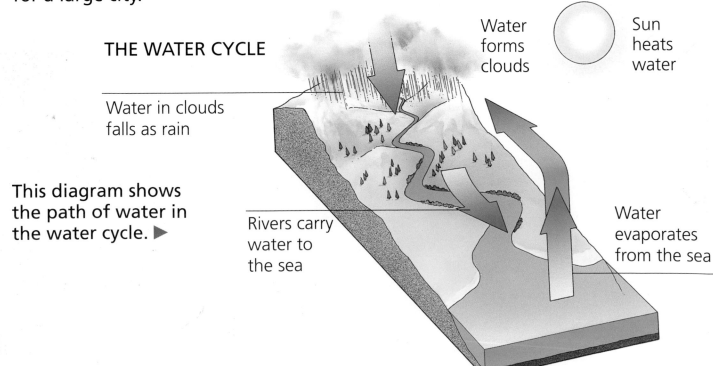

THE WATER CYCLE

Water forms clouds

Sun heats water

Water in clouds falls as rain

This diagram shows the path of water in the water cycle. ▶

Rivers carry water to the sea

Water evaporates from the sea

The Earth from space.
The Earth is the only
planet in the solar
system that has
oceans and rivers.

FACTFILE

Plants are part of the
water cycle. Their roots
pull up water from the
soil to their leaves. It
evaporates from their
leaves into the Earth's
atmosphere.

Plants and animals

All living things need water to survive. The driest places in the world have the fewest plants and animals.

Green plants make their food by mixing water with nutrients in the soil and carbon dioxide. All animals rely on plants for food. They either eat plants themselves, or they eat other animals that eat plants.

Plants grow only where they find water. In Israel's Negev Desert, plants grow along the edge of a stream. ▼

People and water

Over half our body weight is water. We constantly lose water every day through sweating, breathing and going to the toilet. So we need to replace about 2.5 litres through drinking and eating.

We also use water for washing and cooking. Farmers use it to irrigate their crops, and factories use tonnes of water every day.

FACTFILE

Only 3 per cent of the world's water is fresh. The rest is salty. Most freshwater is frozen in ice caps and glaciers. This means we can use only 1 per cent of the world's water, from rivers, lakes and under the ground.

▲ Water always flows downhill because of gravity. Pressure from pumps can send it uphill.

Tides

The sea-level rises and falls about once every twelve hours. The changes in depth are called the tides. They are mainly caused by the Moon's gravity, which pulls water towards it. Seas and oceans closest to the Moon have high tides.

As the Moon goes around the spinning Earth, it rises later each day. This means that high tides are later each day.

The Sun's gravity also pulls water towards it. So the height of the tides depends on where the Sun, Moon and Earth are. When all three are lined up, the tides are highest.

FACTFILE

Tides change the depth of water in the sea. At low tide, harbours can become mud flats. So ship captains have to know about tide times everywhere they go.

A crescent moon rises over the Mediterranean Sea. ▼

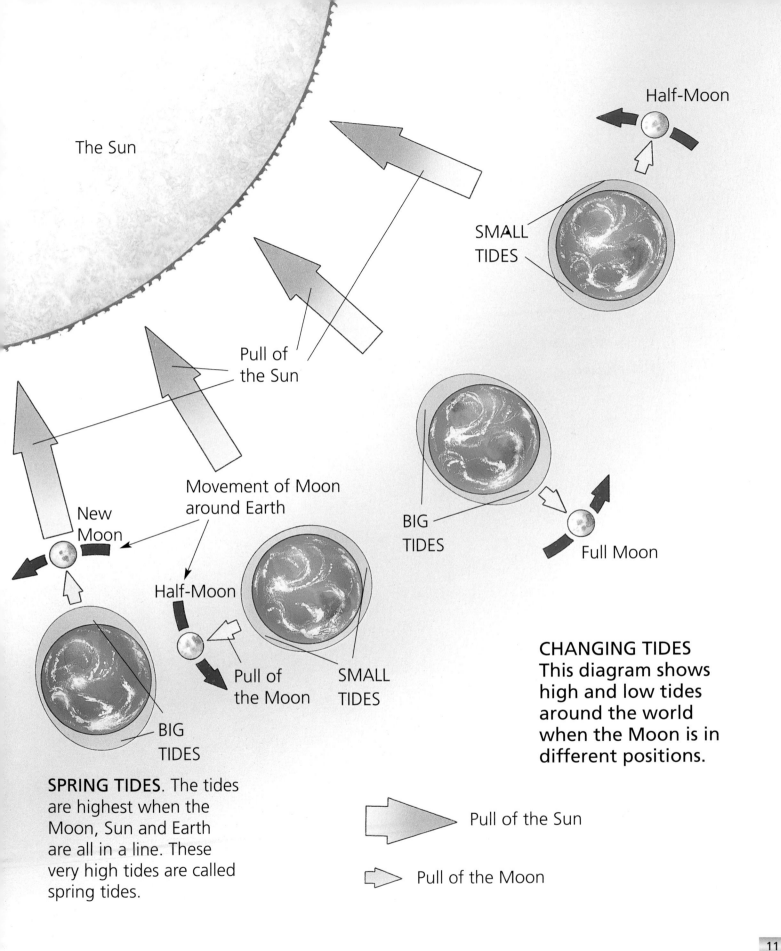

The Sun

Half-Moon

SMALL TIDES

Pull of the Sun

Movement of Moon around Earth

New Moon

BIG TIDES

Full Moon

Half-Moon

Pull of the Moon

SMALL TIDES

Half-Moon

BIG TIDES

SPRING TIDES. The tides are highest when the Moon, Sun and Earth are all in a line. These very high tides are called spring tides.

CHANGING TIDES
This diagram shows high and low tides around the world when the Moon is in different positions.

Pull of the Sun

Pull of the Moon

Erosion and flooding

Water constantly erodes, or wears away, the land. Strong waves and winds wear away coastlines. Beaches are washed away by storms, cliffs fall down, and buildings fall into the sea.

At high tides and during storms, the sea can flood the land. In the Netherlands, flood barriers called dykes were built to protect the low coastlines. In 1953, high tides broke through the dykes and 1,800 people were killed.

Not all floods are dangerous. River floods help farmers because they dump fertile mud on the land around rivers. In countries like Bangladesh, people rely on the rivers flooding every year to keep the land fertile so they can grow crops.

FACTFILE

Massive waves called tsunamis are made under the sea by earthquakes. In 1868, a tsunami hit the coast of Chile. It carried a ship 3 kilometres on to the land.

A house falls into the sea in Ireland because the cliff has been eroded by waves. ▶

Waves have eroded the softer land around these pillars of rock in Cornwall, in England. ▶

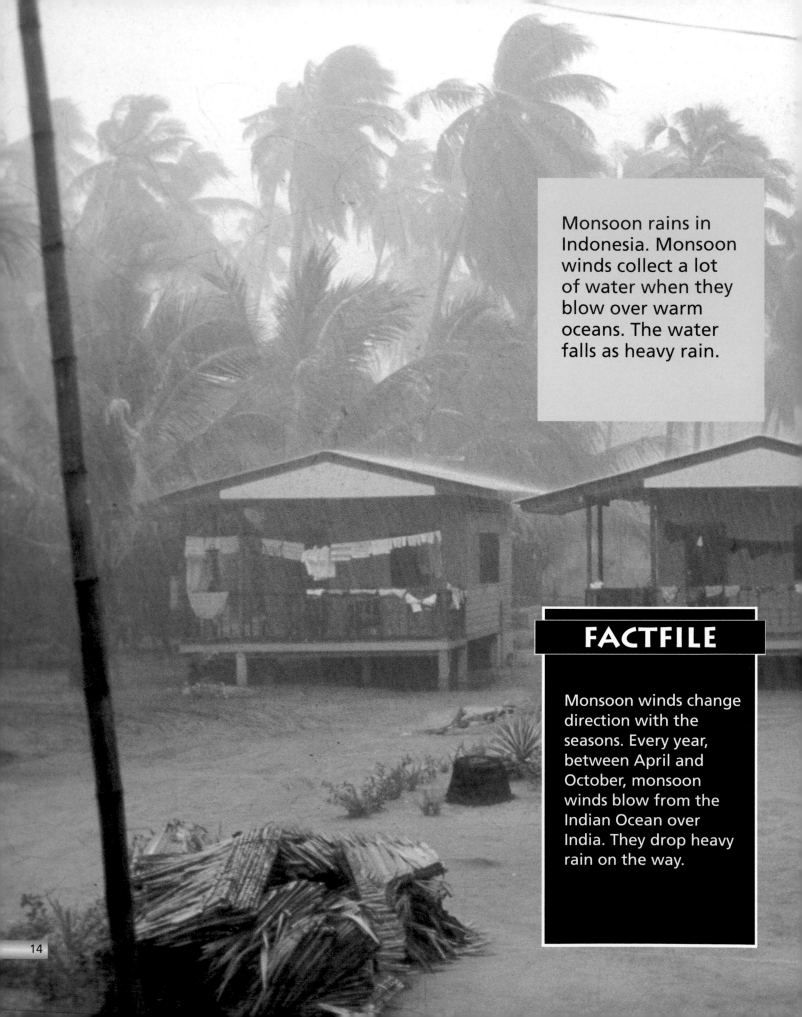

Monsoon rains in Indonesia. Monsoon winds collect a lot of water when they blow over warm oceans. The water falls as heavy rain.

FACTFILE

Monsoon winds change direction with the seasons. Every year, between April and October, monsoon winds blow from the Indian Ocean over India. They drop heavy rain on the way.

Ocean currents and weather

Water in the oceans is constantly on the move. Bodies of water flow in certain directions around the world. These are called ocean currents. They are caused by winds and tides, and by rivers flowing into the sea.

Water also moves up and down in the oceans because water rises when it gets hotter and sinks when it gets cooler. The water's heat affects the weather. When hot water heats air above it, the hot air rises. Later it condenses and forms clouds. Then it falls as rain.

This map shows the path of the main ocean current. It is called the Great Ocean Conveyor Belt. It continually moves all around the world, from pole to pole.

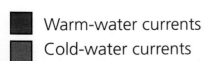

Warm-water currents
Cold-water currents

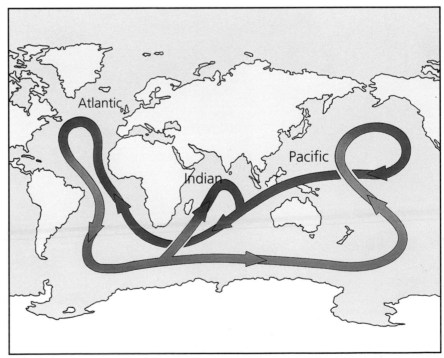

HISTORY OF WATER POWER

Water wheels

Water wheels were the first machines to use water power. They were probably first used by the ancient Egyptians on the River Nile, about 4,000 years ago.

By the year 27 BC, they were in everyday use by the Romans. Most water wheels were used to grind flour, using millstones. Some worked hammers or bellows in forges.

FACTFILE

By the 6th century AD, mills driven by water wheels were used throughout Europe and the Middle East. Growing cities depended on water power to help feed their people. In 1086 there were 5,624 water mills in England. This was written in the Domesday Book.

An old water wheel on the River Test, in England. Only a few mills with working water wheels have survived until modern times.

A French water wheel in the sixteenth century. It works four bellows, which blow air into furnaces to keep them burning. ▶

A water wheel is a wheel with paddles round the edge. When water pushes against the paddles, they turn the wheel and the axle in the centre. The axle could be fixed to millstones. Or it could be fixed to other machines through sets of links.

During the Industrial Revolution, large water wheels were used to run machinery in factories.

Tidal mills

In the estuaries of rivers, tides flow in and out. In the 1100s, engineers found out how to use the rise and fall of the tides for energy.

Wooden gates were built across a river, close to the estuary. When the tides came in, the gates were open.

FACTFILE

In the 1100s, tidal mills were built on the River Adour, in France, and in the estuary of the River Deben, in Suffolk, in England. By the 19th century, there were about 100 tidal mills in Europe.

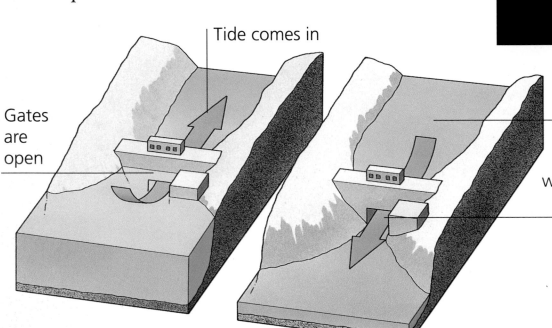

Tide comes in

Gates are open

Water trapped behind gates

Gates let water out through the mill wheels

But when the tides turned to go out, the closed gates trapped the water behind them. Once enough water had collected, it was allowed to flow back through a mill wheel. The collected water flowed much faster than normal, because it was under pressure.

This diagram shows the parts of a windmill and water wheel. It is from a science book called *Physics in Pictures*, published in 1882. ▶

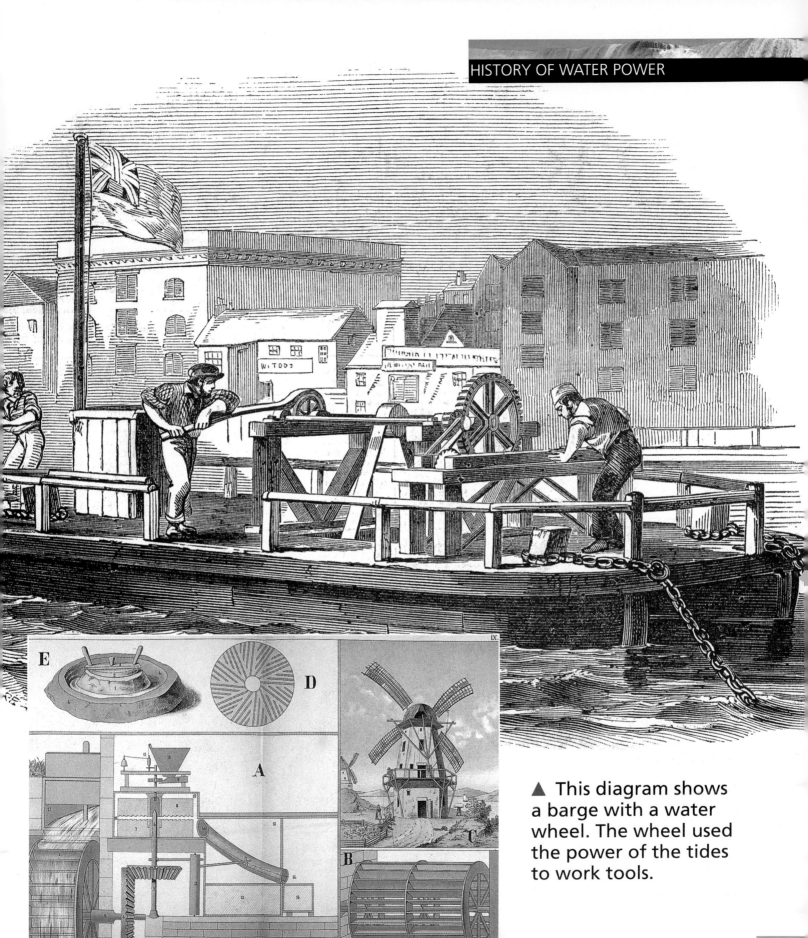

▲ This diagram shows a barge with a water wheel. The wheel used the power of the tides to work tools.

Steam engines

In the eighteenth century, steam engines were used instead of water wheels. The steam engine also used water for power, but it heated up the water first to turn it into steam.

When water turns into steam, it expands quickly. In steam engines, the steam pushed a piston that turned a wheel. The water was heated up by burning coal in a furnace.

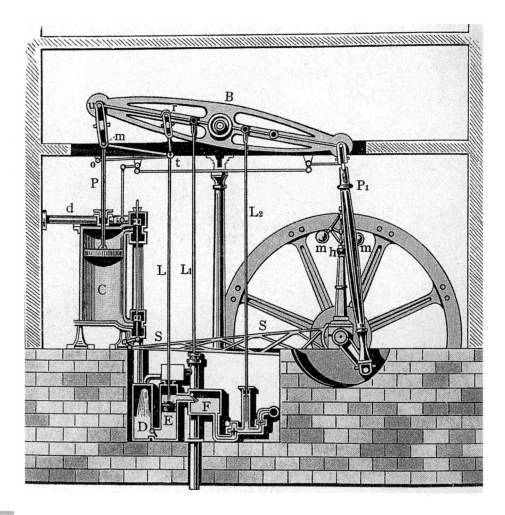

◄ A diagram of the steam engine made by James Watt, in 1765.

Steam pushed a piston down. The piston turned the wheel of a machine. Then the steam was let out. ►

FACTFILE

The first steam locomotive was built in Wales by Richard Trevithick, in 1804. It travelled at a speed of 15 kilometres an hour.

Steam engines started the Industrial Revolution in Europe and the USA.

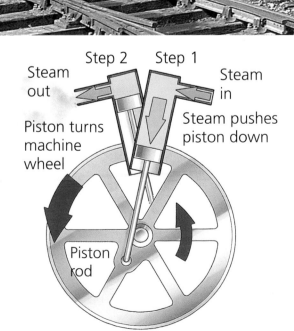

Step 2   Step 1

Steam out

Steam in

Piston turns machine wheel

Steam pushes piston down

Piston rod

The first practical steam engine was invented by Thomas Newcomen, in 1712. It was made to pump water out of deep coal mines.

Then, in 1765, James Watt improved the first steam engine. He used it to drive other machines. Soon there were hundreds of steam engines driving machines in European factories.

USING WATER POWER

Water runs on to an overshot water wheel. An overshot wheel moves three times faster than an undershot wheel of the same size.

FACTFILE

In ancient times, wheels were used to collect water from rivers using pots fixed to their rims. Oxen drove the wheel round. The pots picked up water and emptied it on to the riverbank.

Water wheels

Water wheels can be vertical or horizontal. Vertical wheels are the oldest type. All water wheels have paddles. The water pushes against the paddles to turn the wheel.

There are two types of vertical water wheel: undershot and overshot. Undershot wheels are dipped in water. Their paddles are pushed round by the water below them.

Overshot wheels are pushed round by water running on top of them. They are faster than undershot wheels because they use the weight of the water, as well as its speed.

Horizontal water wheels lie flat. Water is pointed at the paddles through a pipe. The wheel spins round a vertical axis.

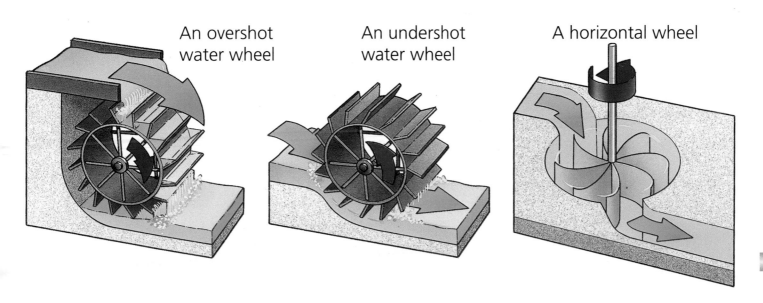

An overshot water wheel

An undershot water wheel

A horizontal wheel

Hydroelectric power

Water power is made into electricity in hydroelectric power stations. To make electricity, water flows through a turbine, making it spin. The turbine drives a generator. The generator makes electricity.

To make the turbine spin as fast as possible and make the most electricity, the water has to be put under pressure. Water pressure is greater in deeper water. So deep reservoirs are made at hydroelectric power stations by building a dam across a river.

The water that flows through the turbine comes from the bottom of the reservoir. It is under the greatest pressure.

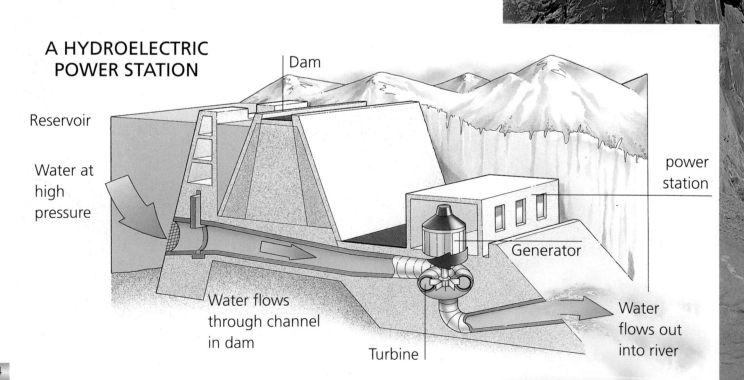

A HYDROELECTRIC POWER STATION

Dam

Reservoir

Water at high pressure

power station

Generator

Water flows through channel in dam

Turbine

Water flows out into river

The Glen Canyon Dam across the Colorado river in Arizona, in the USA. The reservoir crosses the border between Arizona and Utah.

FACTFILE

Hydroelectric power stations make about one-fifth of the world's electricity. The most powerful hydroelectric power station is the Itaipu power station on the Parana river in South America. It makes enough power for several cities.

Pumped-storage power stations

The need for electricity changes every minute. More electricity is needed during the day, when people are awake, than at night. So power stations constantly have to adjust the amount of electricity they supply.

Pumped-storage power stations can store energy. They use two reservoirs. At night, when the need for electricity is low, water is pumped uphill from a lower reservoir to an upper reservoir.

During the day, if there is a need for more electricity, the water from the upper reservoir can be quickly used to make electricity. It falls to the lower reservoir through the turbines.

A hydroelectric power station in Brazil. The pipes in the front of the photo carry water uphill, where it is stored in an upper reservoir. ▼

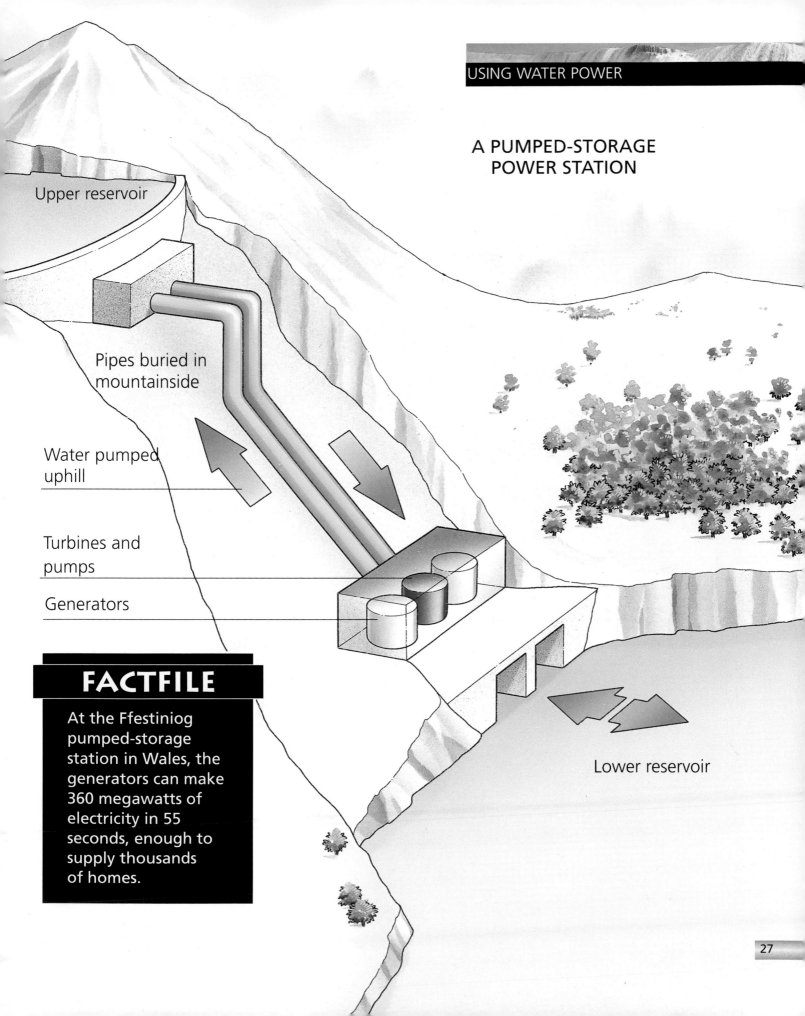

A PUMPED-STORAGE POWER STATION

Upper reservoir

Pipes buried in mountainside

Water pumped uphill

Turbines and pumps

Generators

Lower reservoir

FACTFILE

At the Ffestiniog pumped-storage station in Wales, the generators can make 360 megawatts of electricity in 55 seconds, enough to supply thousands of homes.

People in Norway use more electricity per person than in any other country. Most electricity is used for heating and lighting.

Over 99 per cent of Norway's electricity is made from water power. Instead of building coal or oil-fired power stations, they built hydroelectric power stations instead. The high mountains and heavy rainfall in Norway are ideal for making hydroelectric power.

FACTFILE

Each person in Norway uses about 30,000 kilowatt-hours of electricity a year. In Denmark each person uses about 6,000 kilowatt-hours, and in Switzerland they use only 800 kilowatt-hours each.

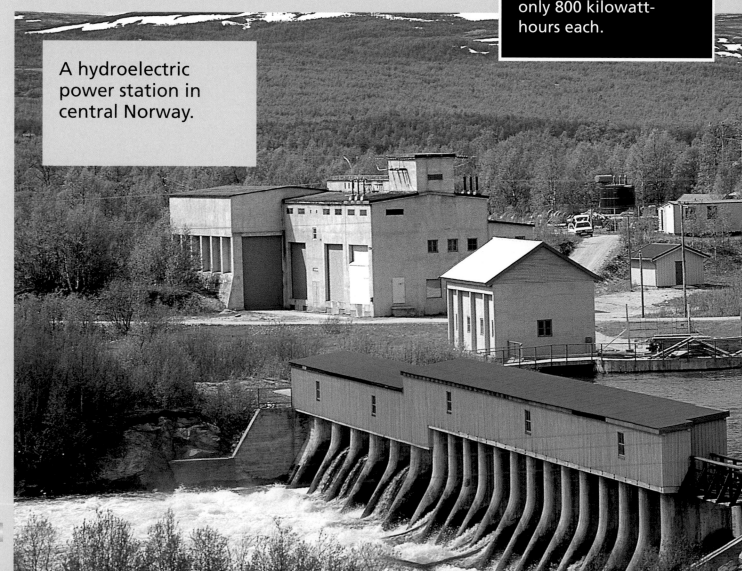

A hydroelectric power station in central Norway.

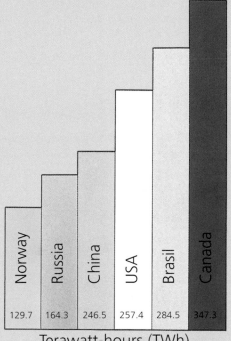

◄ The generator room at a power station in Norway.

Norway	Russia	China	USA	Brasil	Canada
129.7	164.3	246.5	257.4	284.5	347.3

Terawatt-hours (TWh)

This chart shows the six countries that made the most hydroelecticity in 2004. One terawatt hour is a million million watt hours

No other country makes as much of its electricity from water power as Norway. Electricity in Norway is cheap and clean. The only fuel that is burned at a hydroelectric power station is in the machines that built it. This means that the energy used in Norway does less damage to the environment than energy used in countries nearby.

Electricity from waves

The wind whips up the surface of the sea into waves and blows them towards the land. As each wave rolls over the sea, the water level goes up and down. The up-down movement can be used to make electricity with a wave generator.

In a wave generator, a float on the water's surface bobs up and down. It changes the up-down movement into a spinning movement. The spin drives an electricity generator.

A row of wave generators called Salter's Ducks on the surface of the sea. ▼

FACTFILE

Winds are caused by the Sun heating the Earth unevenly. When air above hot land rises, it is replaced by air from cooler areas. These movements of air are the winds.

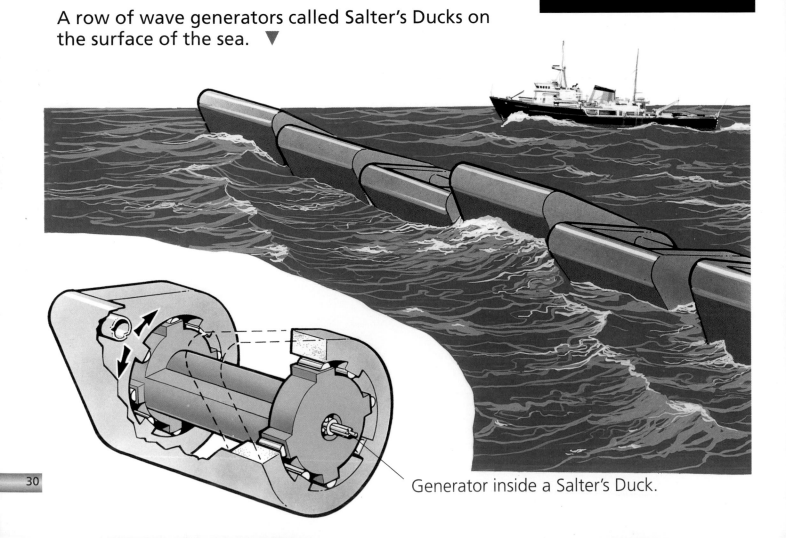

Generator inside a Salter's Duck.

A design for a new wave generator. The up-down movement of the front flaps pumps air through a turbine and electricity generator. ▼

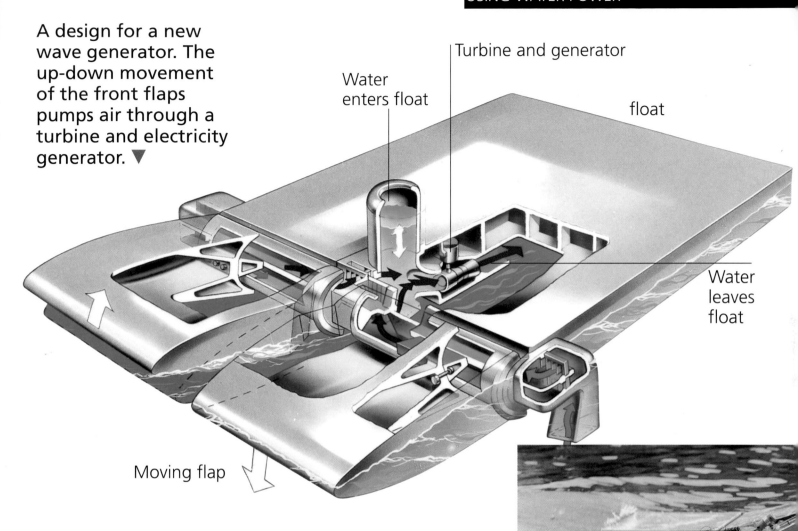

Turbine and generator

Water enters float

float

Water leaves float

Moving flap

Wave generators

Salter's Ducks (see the opposite page) are a type of wave generator. Another wave generator is called a clam.

A clam has a set of floating air bags. The waves squeeze air between the bags as they rise and fall underneath them. The air spins through generators between the bags.

Salter's Ducks and clams are new inventions. They make very little electricity.

▲ Small models of wave generators are tested in a tank of water.

A new machine to use the power of waves is called an oscillating water column generator. It is a vertical pipe fixed to a cliff-face, with an open top and bottom.

The bottom of the pipe sits below the water's surface. Inside the pipe there is a turbine linked to an electricity generator.

Waves force water through a natural blow-hole. ▼

When waves reach the cliff, the water level inside the pipe rises and falls. Air rushes up and down the pipe, making the turbine spin. The turbine drives the electricity generator.

The only water-column generators that have been made so far have been destroyed by battering waves. Engineers are trying to improve its design, and find other ways of using the power of waves.

▲ A water-column generator in Norway.

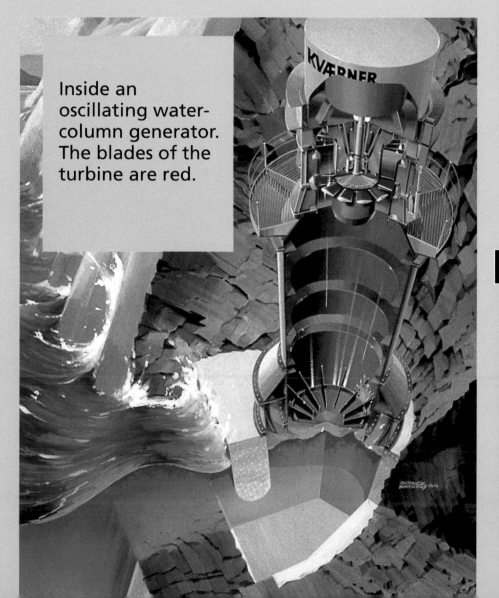

Inside an oscillating water-column generator. The blades of the turbine are red.

FACTFILE

The oscillating water-column generator was first made in the early 1980s. It was invented by student engineers from Northern Ireland and a Norwegian company in Oslo.

Tidal barrages

Modern tidal barriers use the tides in a river estuary just like old tidal mills. A dam or barrage is built across a river. When the tide comes in, the gates of the barrage are open.

At high tide, the gates close and water collects behind the barrage. When the water level in front of the barrage is low, the gates let water rush out through turbines.

Environmental damage

River estuaries are important breeding and feeding grounds for fish, birds and other wildlife. Migrating birds use them as essential feeding stops on their journeys. Tidal barriers can destroy this habitat and its wildlife.

A tidal barrage across a river estuary in the Netherlands. ▼

A curlew sandpiper in a river estuary. Many birds are in danger because tidal barrages can destroy their habitat.

The biggest tidal power station in the world is on the River Rance, in France. It was built in 1966 and makes enough electricity for a quarter of a million homes.

The power station's barrier stretches 750 metres across the river, and the water passes through 24 tunnels. Each tunnel has a turbine and a generator in it.

FACTFILE

In the estuary of the River Rance, the difference in height between high and low tide can be 13.5 metres.

The barrage across the River Rance is also a road.

CUTAWAY DIAGRAM OF THE TIDAL BARRAGE ON THE RIVER RANCE

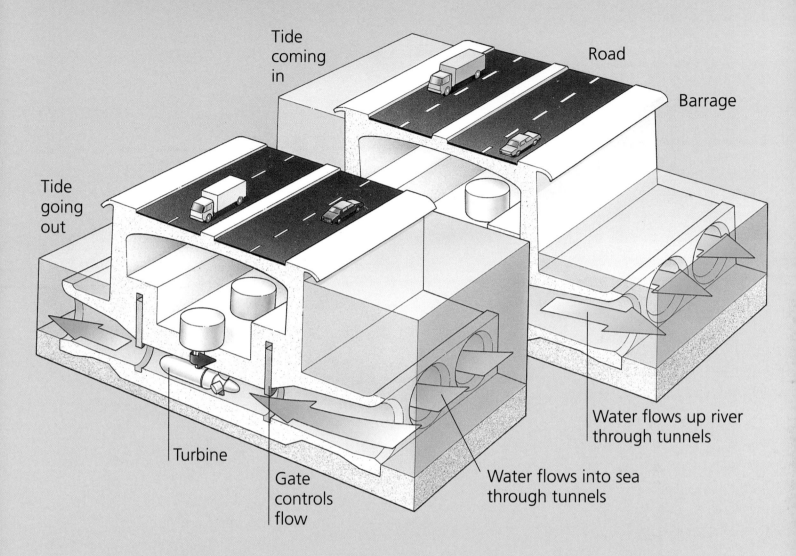

Tide coming in

Road

Barrage

Tide going out

Turbine

Gate controls flow

Water flows into sea through tunnels

Water flows up river through tunnels

▲ The tidal barrage across the River Rance is an important transport route, as well as an electricity generator.

In most tidal power stations, the incoming tide is too slow to turn the turbines.

The River Rance meets the sea in a bay. This funnels the tides coming from the Atlantic up the river mouth. This means the incoming tides are more powerful than most rivers. So the turbines in the power station are driven by the incoming and the outgoing tides.

TURBINES AND ENGINES

Steam turbines are the most powerful type of machinery driven by water today. Like steam engines, steam turbines use the force made when water turns into steam and expands quickly. But they work better than steam engines, because they immediately make a spinning movement instead of using a set of links.

The force of the steam pushes the blades of a turbine inside an air-tight case. The turbine blades turn a shaft. The shaft is linked to an electricity generator.

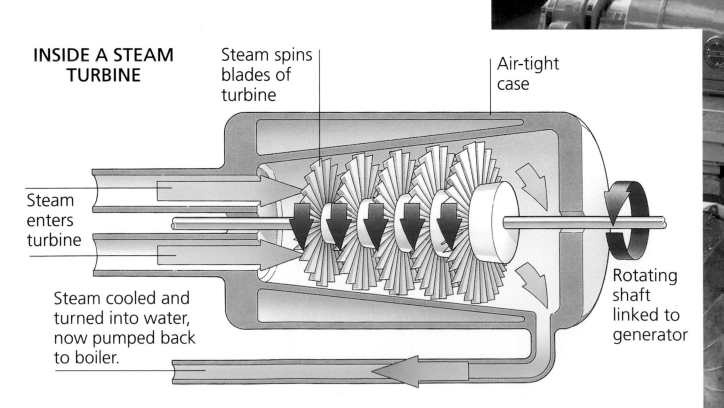

INSIDE A STEAM TURBINE

Steam spins blades of turbine

Air-tight case

Steam enters turbine

Steam cooled and turned into water, now pumped back to boiler.

Rotating shaft linked to generator

FACTFILE

Simple steam turbines use the force of steam on the turbine blades. Other turbines use the speed of the steam as it expands. These turbines have two sets of blades which the steam rushes through at high speed.

Pairs of steam turbines and generators, fed by steam pipes, inside a nuclear power station. ▼

▲ The Boeing Jetfoil is a hydrofoil with water-jet engines.

Water-jet engines

Water-jet engines propel boats along by pumping out a high-speed jet of water. Water is sucked in at the front of the boat. It is passed through a set of high-pressure pumps.

The pumps push the water into a narrow, fast jet. The jet shoots out of the back of the boat, propelling it along.

Car ferries, high-speed boats and sports craft use water-jet engines. Because they do not have propellers, they have fewer problems with shallow water, weeds or ropes.

Water-jet engines are the same as the ones used by fighter-planes. But instead of making a jet of air, they pump out a jet of water.

A ferry such as a SeaCat uses water jets to help it move. ▼

A jet-ski uses a water-jet engine to propel it across the water. ▼

Pumps

Water jet out

Water in

FACTFILE

Scallops are a type of shellfish that use natural jet propulsion to move through the water. They squirt out a jet of water by snapping their shells shut.

Water jets in Alaska wash out gravels that may contain gold.

FACTFILE

Hoses have been used in mining for over a hundred years. But using water jets to cut through materials was only invented in 1968, by Norman Franz in the USA.

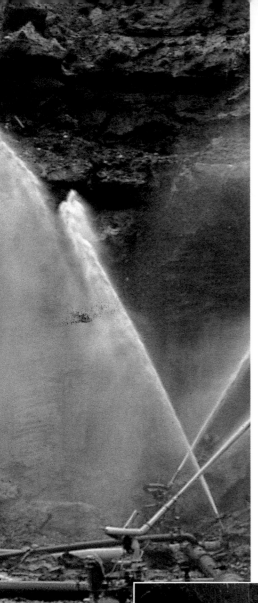

Water-jet tools

When water is pushed into a jet under pressure, it can clean materials by blasting off the dirt. At a higher pressure, it can even cut through hard materials like rock.

Water is put under pressure using a machine called a compressor. It is sprayed out through a hose, with a nozzle that pushes it into a jet.

Water jets are used in mining. They cut through the earth to find minerals. Then they wash the minerals into an area where they can be collected. Water jets are also used to clean buildings that have been blackened by air pollution.

A water-jet tool is used to clean a building.

THE FUTURE OF WATER POWER

As fossil fuels run out, many countries want to make their energy from water and other renewable energy sources. Between 1997 and 2020, the world is expected to use 54 per cent more energy from renewable sources.

FACTFILE

A new machine called an Ocean Thermal Energy Converter is being developed to make electricity in tropical seas. It uses the difference in temperature between different depths of water to drive a turbine.

The Ocean Thermal Energy Converter uses warm water to change a liquid into a gas. The gas drives a turbine. ▶

Warm water goes in

Warm water comes out and rises

Cold water comes out and sinks

Turbine

Cold water rises into the machine and cools the gas

Most new hydroelectric power stations are now being built in the poorer countries of the world, especially Asia. In China, the massive Three Gorges Dam project began in 1993. It should be ready in 2009. Many smaller projects are also being built around the world.

Wave power is a fairly new source of energy. A lot of research is needed to use it as a practical source for the future.

This diagram shows how water and wind power could be used in the future. ▼

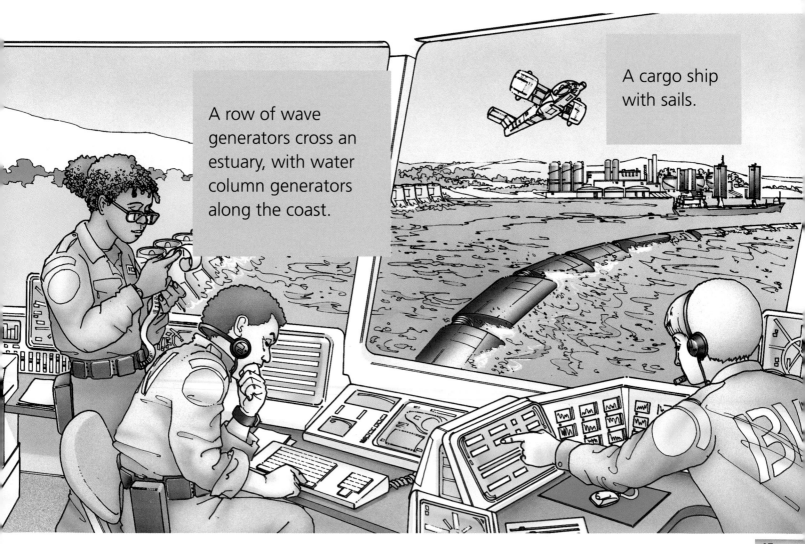

A row of wave generators cross an estuary, with water column generators along the coast.

A cargo ship with sails.

GLOSSARY

Axle A rod that a wheel is fixed to.

Bellows A pair of airbags with handles used for blowing air into a fire.

Blow-hole A hole in rocks or a tunnel through which air, smoke or water can escape.

Cargo The load of goods carried by a ship or an aircraft.

Condenses Changes from a gas into a liquid by cooling.

Currents Bodies of water moving in a certain direction.

Domesday Book The record of the lands of England made in 1086 by order of William I.

Engineers People who design and build things such as machines and bridges.

Erosion The wearing away of something, like land.

Estuary A broad mouth of a river, where tides flow in and out.

Evaporates Turns into a gas from a liquid or solid.

Expand To get bigger.

Forges Workshops were metals were heated and shaped.

Fossil fuels Coal, oil and natural gas formed from the remains of plants and animals over millions of years.

Furnace A container for a very hot fire.

Generator A machine that changes movement energy into electricity.

Glacier A large mass of ice that moves very slowly down a mountain or valley.

Gravity The force that attracts objects towards the centre of the Earth.

Habitats Natural homes of plants and animals.

Horizontal Parallel to the horizon.

Hydroelectric Energy made from the power of fast-flowing water.

Irrigation The supply of water to land.

Kilowatt-hours (kWh) A unit of energy that is equal to 1,000 watts of electrical power being used for one hour.

Millstones A pair of rounded, flat stones used for grinding corn, wheat or other grain.

Propel To drive or push forward. Propulsion is the act of driving something forward.

Renewable Something that can be replaced.

Reservoir A place where water is collected and stored.

Rotating Turning around a centre, such as a shaft.

Shaft A bar that helps parts of a machine turn.

Turbines Angled blades fitted to a shaft that are made to rotate by the force of water, steam or air.

Vertical Straight up and down.

Wave generators Device that makes electricity from waves using floats and electricity generators.

FURTHER INFORMATION

Books to read

Action for the Environment: Energy Supplies by Chris Oxlade and Rufus Bellamy (Franklin Watts, 2004)

Alpha Science: Energy by Sally Morgan (Evans, 1997)

A Closer Look at the Greenhouse Effect by Alex Edmonds (Franklin Watts, 1999)

Cycles in Science: Energy by Peter D. Riley (Heinemann, 1997)

Essential Energy: Energy Alternatives by Robert Snedden (Heinemann, 2002)

Future Tech: Energy by Sally Morgan (Belitha, 1999)

Saving Our World: New Energy Sources by N. Hawkes (Franklin Watts, 2003)

Science Topics: Energy by Chris Oxlade (Heinemann, 1998)

Step-by-Step Science: Energy and Movement by Chris Oxlade (Franklin Watts, 2002)

Sustainable World: Energy by Rob Bowden (Hodder Wayland, 2003)

Power station produces several million watts.

Family house uses a few thousand watts.

Washing machine: 2,500 watts

Electric iron: 1,000 watts

Light bulb: 100 watts

ENERGY CONSUMPTION

The use of energy is measured in joules per second, or watts. Different machines use up different amounts of energy. The diagram on the right gives a few examples. ▶

INDEX